MARS Railroad,
a STEM Design Exercise

by Patrick H. Stakem

(c) 2018

Number 7 in the STEM Series

Table of Contents

Introduction

I put this out to the world-wide STEM community as a challenge. Design a viable railroad for Mars. We are going to need it eventually. It will take a while. Start now.

When humans arrive on Mars, and for a while afterwards, they will be completely dependent on periodic resupply missions from Earth. They will need to find their own supplies, chiefly water and methane, later, minerals and metals. Mars will be in the pre-Industrial Revolution phase, Just like when the early American settlers were dependent on resupply ships from Europe. Except the early American settlers found abundant food and indigenous people.

Railroading on Mars. This has not yet been done. It will be done eventually by someone. Why not you?

Author

The author has a BSEE in Electrical Engineering from Carnegie-Mellon University, and Masters Degrees in Applied Physics and Computer Science from the Johns Hopkins University. During a career as a NASA support contractor from 1971 to 2013, he worked at all of the NASA Centers. He served as a mentor for the NASA/GSFC Summer Robotics Engineering Boot Camp at GSFC for 2 years. He teaches Embedded Systems for the Johns Hopkins University, Engineering for Professionals Program, and has done several summer Cubesat Programs at the undergraduate and graduate level. He supports International collaborative open-source projects, and STEM Programs.

He also served as Vice-President of Technology for a short-line railroad.

Mr. Stakem has been affiliated with the Whiting School of Engineering of the Johns Hopkins University.

What is STEM?

STEM (Science, Technology, Engineering, Mathematics) is the key to the United States' continued dominance in High Technology. It took a lot of expertise to implement the first cell phone. Now they are turned out like cookies in third world countries. Let's raise all country's tech savvy by hands-on, mind-stretching exercises.

STEM addresses overall education policy and curriculum sources in schools, at critical grade levels.

Although the teachers are experts in their particular area, and know how to present grade-appropriate material, they may not necessarily know how to find and access advanced resources they need, or where to get help in a particular topic area.

STEM programs are seen as critically important in the education system, world-wide. It is getting to be a complex, interconnected ecosystem. Advances in the subject areas of STEM will take place only by those who know how to exploit this ecosystem for knowledge. It is available on the Internet.

When I was in school well before the Internet and STEM age, I had an encyclopedia, updated yearly. Today, students can access WikiPedia from their smart phones. The focus has changed from knowing facts, which are at your fingertips, on demand, to leveraging facts to innovate. This approach touches all of the academic disciplines, the Humanities, Languages, Art, besides the traditional STEM topics. Perhaps the best skill set to have is good internet search skills. Teachers have had to transition from asking factual questions, to asking questions that derive from applications of online research, and accrued knowledge.

When I was a kid, there was no STEM. My interests in science and engineering led to research and hands-on experimentation. Luckily, I survived. I was called on, while in grade school, to demonstrate some concepts of electricity to a High School class. The first satellite was launched, and I was glued to the black & white TV. I participated in Model Rocketry at the High School Level, and went on to fly Nationally. This was made possible by an extraordinary

High School Science teacher. I made quite a few friends, some of whom became Astronauts. I was given a great opportunity when I received a full scholarship to a College of my choice. I went to Carnegie Tech in Pittsburgh (now, Carnegie-Mellon University), and launched a career in Engineering and Aerospace. It is time for me to pay forward.

I think that he Mars Railroad can be a major focal point for STEM, embracing a wide variety of topics at the cutting edge of technology and science. I have a handful of technical degrees, and spent 42 years at the various NASA Centers. I teach Electrical Engineering courses world-wide, and have done specialty Cubesat courses at the undergraduate and graduate level. It is time to apply that expertise earlier in the education process.

My thesis is, a project brings together all of the interesting pieces to provide a focal point for student work. There is a massive body of applicable free support material available. I have experience teaching engineering courses at the advanced undergraduate and graduate level, but I have no experience or credentials at the critical pre-K thru 12 levels.

I think STEM is a critical resource for understanding and implementing the future. Let's do this. Future generations of STEM-mies will thank us.

Although STEM schools will have in-house expertise in Science (Physics, Chemistry, Biology), and Math (counting numbers through calculus), they are not heavily into Technology and Engineering. That's where you can ask for help – there's a lot of resources and knowledgeable individuals available out there.

A note on Units

I am fairly conversant in both English and Metric units (what is the metric equivalent of furlongs per fortnight?). Metric (SI) is mandated for NASA usage now, for interchangeability with our partner space faring nations. When a lot of the legacy space flights discussed here were flown, English units were the norm. I have tried to keep the units comparable to the mission at the time. Conversions are easy enough, but units conversion is a source of error. It has caused loss of missions and lives. It's not what you know about units and measurement, its how you think. And, I still think English units (even the English use Metric now), and convert in my head or on my phone.

For scientific/engineering work, the Metric system is well thought out. For artisans, the English system served well, as most units were divisible by 2, which is easy. Fold the cloth. Hopefully, when we are all taught Metric first, some one will still remember the conversions. You just need a good slide rule....

Before we start thinking of the Mars railroad, we should consider the railroad in space, currently operating. That is a 278 foot line on the outside of the International Space Station. NASA says it is both the slowest and fastest railroading the universe. It crawls along at about an inch per second, but the station to which it is attached is moving about 17,000 mph. It is actually called the ISS Mobile Transporter., and is a base platform on rails, for robotic arms. It can transport the CanadaArm, or the robot Dextre. It can also be used to transport an Astronaut.

Getting to Mars

Getting to Mars is not easy, and requires a lot of energy and time. First, we get to Earth orbit, and check that all the systems have survived launch. Knowing where Mars is currently, we can calculate the correct point to head to, when our mission reaches the correct distance. When the mission reaches Mars, we have to slow down and enter Martian orbit. Counter-intuitively, a minimum energy mission (but not a minimum time scenario) takes us looping past Venus to get to Mars in an efficient manner.

The distance from Earth to Mars varies on their relative positions in their orbits, but ranges from 33.9 to 250 million miles.

Every 26 months, the Earth and Mars align such that a minimum energy transfer can get a spacecraft from one to the other. The implications are that if you miss the launch window, you need to wait some 2 years, and that there is a minimum time for a Earth to Mars journey, landing, and return to Earth. This sets a mission duration that is important for future crewed missions. The travel time for this path is about nine months.

A Hohmann transfer orbit is an elliptical orbit between the circular orbits of two planets. It allows for the maximum payload to be sent with a given energy expenditure (or, amount of rocket fuel). You can get there quicker by expending more energy. The alignment, or launch window, is critical. The details of the interplanetary transfer orbit were defined and published by German scientist Walter Hohmann in 1925.

The hope is, we can establish a refueling station on Mars or one of its moons, so we don't need to carry our return fuel with us. Since Mars has been proven to have water ice, this might be quite feasible. Using solar energy, you can break down water into liquid hydrogen and liquid oxygen – rocket fuel and oxydizer.

The Mars Environment

Mars is the next planet out from the Sun, from the Earth. It is almost the smallest, but Mercury has that title. The red of the red planet comes from iron oxides. Having a slight gravity, it has a thin atmosphere. This means that many things evaporate into space, such as water. If water exists on Mars, it is in the form of ice. Indeed, the Martian poles have water ice. There was a large volume of underground ice discovered in the area of Utopia Planitia. Mars can be seen by the naked eye, from Earth's surface. Mars scores the largest the largest known volcano in the solar system, Olympus Mons. It has two moons. It's spin axes is tilted like the Earth's, so it has seasons. It consists mostly of desert terrain, with tall mountains and impact craters. No oceans, no canals.

The one-way light time between Earth and Mars varies between 3 and 29 minutes, depending on the relative positions in their orbits. This affects the communication between the Earth, and the various orbiters, landers, and rovers. In some cases, when Mars and the Earth are on opposite sides of the Sun, communication is not possible. Optical telescopes on Earth can resolve features down to a size of 300km, due to the distortions of Earth's atmosphere. This drives the many imaging missions that have visited the red planet.

Mars' size is about half of Earth's, but it has about the same dry-land surface area. It has a metallic core, but not a significant magnetic field. We have good data on the composition of the surface courtesy of the Phoenix lander. Surface features from a variety of sources seem to show erosion. More water certainly existed on Mars in earlier periods. The poles are very cold, and covered in solid carbon dioxide. The United States Geologic Survey maintains accurate maps of the surface. Mars has a large number of impact craters, due to the thin atmosphere.

Mars has polar ice caps, of frozen water ice, and sections of carbon dioxide. When the poles are in Sunlight, this "snow" or ice sublimates. This action results in clouds, and geysers of CO2 gas. Mars gets less than half the amount of sunlight as does the Earth. When Mars is closest to the Sun, large dust clouds are generated, that can envelope the planet. One of these happened in 2018, affecting the operation of surface rovers. Mars has an observing weather satellite in orbit.

The Martian surface is blasted by the solar wind, due to its thin atmosphere, and lack of a magnetic field. Most of the Martian atmosphere is carbon dioxide. Methane is present, suggesting a biological origin, but other sources are possible. The current Indian Mars mission is specifically searching for methane in the atmosphere. Mars has awesome dust storms, some that cover the entire planet.

Due to its distance from the Sun, Mars only receives about 45% of the solar energy that the Earth gets. It can warm up on the surface during the day, but it always plummets at night, usually to where the carbon dioxide in the atmosphere freezes. Wind speeds can get up to 60 mph or so, but the atmosphere is so thin, you many not notice.

Mars was one of the items of interest for early astronomers. The Greeks and Romans named it after their god of war. The Sumerians called it Nergal, after their god of war. In Niveveh, the cult of the red planet called it "the star of judgment of the fate of the dead." The Egyptians were well aware of the planet, and the strange path it took across the heavens. Aristotle noted that Mars disappeared behind the moon, proving it was further away. Galileo was the first to study Mars with a telescope. The Chinese called it the fire star.

If you look in the Mayan document called the Dresden Codex, you will see an extensive list of observations of Mars, what we now call an Ephemeris.

Asteroids, chunks of rock from Mars, are usually found in Antarctica. My old college college professor figured out why, and wrote the definitive paper. I haven't a clue.

One thing that would make Mars more habitable for humans has been very bad for Venus, and is a problem for Earth. Greenhouse gases on Mars would add density to the atmosphere, and control the wide swings of temperature. Not quite ready for terraforming.

Appropriate technology

This section discusses how to identify and apply appropriate technology to the unique problem set, Mars railroads.

NASA, like many large endeavors, has a defined engineering approach for design, develop, and testing. This is outlined in the NASA System Engineering Handbook, which can be downloaded for free. This manual defines processes that, historically, work. It is applicable to both hardware and software. One of the more useful concepts is the Technology Readiness Level.

The Technology readiness level (TRL) is a measure of a device's maturity for use. There are different TRL definitions by different agencies (NASA, DoD, ESA, FAA, DOE, etc). TRL are based on a scale from 1 to 9 with 9 being the most mature technology. The use of TRLs enables consistent, uniform, discussions of technical maturity across different types of technology. We will discuss the NASA one here, which is the original definition from the 1980's.

Technology readiness levels in the National Aeronautics and Space Administration (NASA)

1. Basic principles observed and reported
This is the lowest "level" of technology maturation. At this level, scientific research begins to be translated into applied research and development.

2. Technology concept and/or application formulated
Once basic physical principles are observed, then at the next level of maturation, practical applications of those characteristics can be 'invented' or identified. At this level, the application is still speculative: there is not experimental proof or detailed analysis to support the conjecture.

3. Analytical and experimental critical function and/or characteristic proof of concept.

At this step in the maturation process, active research and development (R&D) is initiated. This must include both analytical studies to set the technology into an appropriate context and laboratory-based studies to physically validate that the analytical predictions are correct. These studies and experiments should constitute "proof-of-concept" validation of the applications/concepts formulated at TRL 2.

4. Component and/or breadboard validation in laboratory environment.

Following successful "proof-of-concept" work, basic technological elements must be integrated to establish that the "pieces" will work together to achieve concept-enabling levels of performance for a component and/or breadboard. This validation must be devised to support the concept that was formulated earlier, and should also be consistent with the requirements of potential system applications. The validation is "low-fidelity" compared to the eventual system: it could be composed of ad hoc discrete components in a laboratory

TRL's can be applied to hardware or software, components, boxes, subsystems, or systems. Ultimately, we want the TRL level for the entire systems to be consistent with our flight requirements. Some components may have higher levels than needed.

5. Component and/or breadboard validation in relevant environment.

At this level, the fidelity of the component and/or breadboard being tested has to increase significantly. The basic technological elements must be integrated with reasonably realistic supporting elements so that the total applications (component-level, sub-system level, or system-level) can be tested in a 'simulated' or somewhat realistic environment.

6. System/subsystem model or prototype demonstration in a relevant environment (ground or space).

A major step in the level of fidelity of the technology demonstration follows the completion of TRL 5. At TRL 6, a representative model or prototype system or system - which would go well beyond ad hoc, 'patch-cord' or discrete component level breadboarding - would be tested in a relevant environment. At this level, if the only 'relevant environment' is the environment of space, then the model/prototype must be demonstrated in space.

7. System prototype demonstration in a space environment.

TRL 7 is a significant step beyond TRL 6, requiring an actual system prototype demonstration in a space environment. The prototype should be near or at the scale of the planned operational system and the demonstration must take place in space.

The TRL assessment allows us to consider the readiness and risk of our technology elements, and of the system.

Wheels on the ground

As of this writing, three Nations have sent missions to Mars: The United States, Russia, China, and India. ESA, consisting of a number of European nations, has also implemented Mars missions. The Emirates have sent a flyby mission

Mars has an orbital infrastructure in place, to support ground-based missions. This includes weather satellites, to track dust storms, and communications relays back to Earth.

As of this writing, there are three payloads headed to the red planet, using the current transfer opportunity.

Sojourner

The first wheeled vehicle on Mars was Sojourner, in 1997. It was a 6-wheeled design, with a solar panel for power, but the batteries were not rechargeable. It lasted from July through September. It had a complex problem in its on-board computer, that was diagnosed and corrected from Earth.

Mars-2 and Mars-3

The Russian Mars-2 failed when its landing module crashed. Mars-3's mission was over about 20 seconds after landing, when communications was lost. These rovers were to use skis.

Mars Exploration Rover

The MER are six-wheeled, 400 pound solar-powered robots, launched in 2003 as part of NASA's ongoing Mars Exploration Program. *Opportunity* (MER-B) landed successfully at Meridiani Planum on Mars on January 25, 2004, three weeks after its twin *Spirit* (MER-A) had landed on the other side of the planet.

For power, they use 140 watt solar arrays and rechargable Li-ion batteries. The Rovers require 100 watts for driving, One problem that was noted was that the Martian dust storms cover the solar panels with fine dust, reducing their efficiency. This resulted in the use of a radioisotope generator on a subsequent mission. It's been observed that Rovers often use more energy in path planning, than to execute the actual path. The advantage of a vehicle on tracks is, it doesn't have to do navigation.

This is an ongoing mission. It was originally planned for 90 days, but the *Opportunity* Rover is still collecting useful data regarding potential life on our sister planet some 11 years later as of this writing. It has traveled over 35 kilometers on the Martian surface.

Curiosity

The Rover vehicle *Curiosity* weights just about 1 ton (2,000 lbs.) and is 10 feet long. It has autonomous navigation over the surface, and is expected to cover about 12 miles over the life of the mission. The platform uses six wheels The Rover Compute Elements are based on the BAE Systems' RAD-750 CPU, rated at 400 mips. Each computer has 256 Mbytes of RAM, and 2 Gbytes of flash memory. The power source for the rover is a radioisotope thermal power system providing both electricity and heat. It is rated at 125 electrical watts, and 2,000 thermal watts, at the beginning of the mission. The operating system is WindRiver's VxWorks real-time operating system. The vehicle was assembled and tested at NASA's Goddard Space Flight Center, and shipped to lead center JPL. The landing location in Gale Crater was named Bradbury Landing, after the science fiction writer, Ray Bradbury.

Mars figured heavily in his writings. Gale Crater is named after an Australian amateur astronomer, Walter Gale. There is some evidence that the crater was once filled with water.

Insight

The Insight (Interior Exploration using Seismic Investigations, Geodesy and Heat Transport) Mission launched in May of 2018, and is scheduled to land on Mars in November of that year. It is a robotic lander with a seismometer and a heat transfer probe. It is based on the earlier Phoenix design.

The spacecraft landed on the Martian surface at Elysium Planitia in November of 2018. This is the latest Mars mssion, but does not include any wheels. The mission is going quite well.

Mars Express

This ESA mission, in 2003, was a success, and remains so as of this writing. A lander, Beagle2, was deployed, but its solar panels failed to deploy, ending its part of the mission.

Exo-Mars Rover

The Exo-Mars mission to the red planet is a joint effort of the European Space Agency, and the Russians. The initial launch placed the trace gas orbiter at Mars in 2016. Unfortunately, the Lander crashed. The second phase of the mission, to be launched in 2020 is to place a Russian-built Rover on the surface.

Mars 2020

NASA's Mars mission involves a 6-wheeled rover for surface exploration. It is to search for biosignatures. The rover has updated wheels over previous missions, due to lessons learned. They are aluminum, and the width was reduced and the diameter (some 20 inches) increased.

Russian Plans

Russia has ambitious plans for Mars manned exploration and colonization. Part of these is the Mars train, first discussed in the 1960's. The core of the plan involved a train of five movable platforms on the surface, one being the crew cabin, another being a platform for aircraft, two platforms for return ships and a backup, with the third platform being a nuclear reactor for power. The train would survey the Martian surface for a year, when the launch window for a return to Earth would open up. This "train" was a wheeled convoy.

The Hope Mars Mission

I want to mention this mission here, although it does not involve a lander. The Hope mission is being developed by The Emirates, not a big player in the space exploration business, just yet. This is slated to reach Mars in 2021. It will study atmosphere and climate. It will also put the Emirates in the fraternity of spacefaring nations. Welcome to Mars.

Technology of Railroads

First, we need to examine railroads on Earth, to determine what might work on Mars, and what needs a re-think.

Elon Musk's announcement of having a pending fleet of spacecraft to bring a lot of humans to Mars sparked interest in discussing the infrastructure of the outpost. He said, "Its a lot like building the Union Pacific Railroad." Well, Union Pacific replied, "We call dibs on building the first Martian railroad."

Desert Railroad

Perhaps the closest railroad to what we might see on Mars is the National Railway of Mauritania. This is a traditional steel track on ties arrangement, and crosses more than 430 miles of the Sahara, from the iron mines at Zouerate (in the interior) to the port of

Nouashibou on the Atlantic. The ambient temperature is around 120 degrees, F. Not in the shade, because there is no shade.

The locomotives are standard units from U. S. Manufacturer EMD, specified modified to operate in a sandy environment. The heat problem is addressed with additional insulation, and cooling systems. Dust and sand infiltration is the big problem. They have multiple sand plows, including one that can be moved by the engineer in the cab. Wind-blown drifts are an issue.

For desert operations, a pulse filtration system was added, as well as enhanced door and car body sealing. The cabin and body shell is pressurized. The gear cases have labyrinth sealing, so they can out-gas, but sand cannot not intrude on the moving parts. The units maintain their dynamic brakes. All fan blades receive protective coating to counter sand erosion.

So, what is applicable to the Mars environment? Sand/dust plows, sealed gear cases, dynamic brakes, coated fan blades.

In the Martian environment, the cab would be accessed via an airlock, and the machinery would be mostly in a sealed environment, pressurized to just above Martian ambient. The big issue is cleaning off all the dust after entering the air lock, so it does not get in the cabin. This might be accomplished by over pressurizing the airlock, and then cycling it to the outside, before the crew member enters the cabin.

Dynamic brakes run the electric drive `motors backwards, generating power as the train slows, and, on Earth, dissipating the heat via large resistor banks. A better approach is to follow the hybrid car scenario, where the energy is stored in batteries. This is being studied for terrestrial locomotives, but a very large battery would be required. The Mars locomotive would likely be a "hybrid" design with a large battery, to allow it to "limp home" is external power is lost.

The Quinghai-Tibet Railroad

This is a very high elevation railway, built by the Chinese in Tibet. There are some lessons learned here, due to the altitude the train operates at. The line is 1,215 miles long, and stretches from Beijing to Lhasa. It is primarily a passenger line that can also carry freight.

The railroad uses the Tanggula Pass, at 16,650 feet in elevation. There are 675 bridges along the route, and many tunnels. Some bridges were installed so as to not interrupt animal migration.

There are oxygen masks for each passenger. There have been deaths on the train. For me, I know I start feeling the altitude beyond 12,000 feet.

Bombardier built 360 of the high altitude passenger cars, incorporating UV protection on the windows, and an enriched oxygen system. The outside air has about 40% less oxygen than at sea level. Oxygen producing facilities were built along the line. Each car collects its own trash, which is returned to China.

Seventy-eight locomotives were built by GE Transportation. The Qishuyan Locomotive Factory supplied similar units. EMD built some locomotives in China for the line. These were dual cab in configuration, meaning they can run either way without turning.

Some of the rail line is built on permafrost, and this tends to thaw in the summer (so much for "perma.") Heat from the train is responsible for some of this. In some areas, the track is elevated, supported with piles driven deep into the ground. Sections of the track are passively cooled with ammonia heat exchangers. There are earthquake sensors along the route.

There were no fatalities listed from the construction of the rail line.

The Trans-Siberian Railway

The Trans-Siberian railway runs from Moscow to the Pacific, some 5,775 miles. It is the longest in the world, and goes through some of the most rugged terrain on the planet. It has been operating since 1916. Construction started in 1891. One of the more interesting sections was across Lake Baikal, on the SS Baikal, and ice-braking train ferry. A spur line now goes around the lake. Originally operated by steam, the line was electrified in 1929. It is built to the Russian gauge (width between rails) of 5 feet. The U.S. Standard is 4 feet, 8.5 inches between the rails.

The U. S. Transcontinental Railroad.

The U.S. Transcontinental linked the Atlantic with the Pacific. This enabled more rapid passenger (emigrant) travel to the West, where the previous option was the Prairie Schooner, a wagon drawn by oxen or horses. There were several transcontinental railroads built across the U. S. and Canada, the first built by the Central Pacific and the Union Pacific. They started from each end, and built towards the middle. Work trains would be pushed to the end of the line, the cars of rail, ties, and spikes available to the workers.

The railroad closed a 1,750 mile gap, previously between Omaha and Sacramento.

The NorthEast Corridor

The United State's North East Corridor is a legacy electrified line that is owned and operated by Amtrak. It extends from Washington, D. C. to Boston. The rail is shared with freight trains in some sections. It is busy, hosting some 2,200 trains a day.

The line started out using steam, and went to electrification in 1905 through 1990. Some sections used electrified 3rd rail, but the main line was developed with overhead catenary. Delays on the line were due to the Great Depression and World War-II.

The part owned by the Pennsylvania Railroad used 11,000 volts, 25 HZ AC power. The reason for 25 Hz power, instead of the U. S. standard 60Hz is obscured by history. It does mean that mechanical systems were needed to use a 60 Hz motor driving a 25 Hz generator at various locations. The current system still uses this power, as the infrastructure and motive power is in place.

Many other nations have electrified train service. Most subways use third rail systems inside tunnels. Overhead DC systems are used in Japan and other eastern countries, as well as some of Europe and Australia. Street cars systems generally use overhead wires. Third rail systems work up to about 1,500 volts. Parts of the Paris Metro uses a 4-rail scheme, with rubber tires in a guideway, and two guidebars for electrical power.

Alternating current has the advantage over direct current in that transformers can raise or lower the voltage. Transmission at high voltages and lower currents mean less line loss. Direct current has an advantage in that it can be stored in batteries. Regenerative braking system can return power to the grid.

Overhead power lines can be damaged by high winds, which are prevalent on Mars. Blowing sand in the high winds can bury track and damage moving parts.

Outline of Martian Railroads

This sections discusses some constraints of Martian railroads, due to the difference in environments, and how these might be addressed.

How do you build a railroad?

A locomotive is rated by it's tractive effort, a function of its weight, and the friction between the rails and the wheels. On Mars, what we have working against us is the gravity is 38% of Earth's. Thus, for an equivalent weight engine, we only only get 38% of the tractive power that we would on Earth.

Locomotives in common on Earth use regenerative braking in lieu of actual friction brakes where possible. This is usually implemented in a system that allows the train, when going downhill, to run the motors backwards, acting as generators. Then the generated electricity is dissipated as heat. Progress is being made in equipping those locomotives with batteries, to capture the energy. That's like letting your hybrid automobile free roll down hill to charge the batteries. I tried at the east exit from Yellowstone Park and got an effective 74 mpg.

Motive power

The Martian railroad would need motive power, and there are multiple approaches to this. There is a certain nostalgia for steam engines, but these are problematic on Mars. First, we lack oxygen to permit combustion. We could have plenty of methane fuel, but the oxygen is, at the level of current knowledge, no present in enough volume to be useful. Methane, essentially Liquid Natural Gas (LNG) can be stored as a liquid, at -260 degrees. You may have a can of LNG for your barbecue grill.

The engine, should we be able to build it, would probably be a condensing, closed cycle format, to preserve precious water.

A similar problem occurs with the usage of an internal combustion engine, such as a diesel or gas engine. An off-the-shelf methane internal combustion engine could be used, but the oxygen source is missing. The first use for oxygen found or produced on Mars will be to support human habitation. This is likely to be in small batches, from greenhouses. There is methane in the Martian atmosphere

Ok, so what about electric power. Mars is further from the Sun from the Earth, and solar cells are less effective. We might consider wind turbines, but the maximum Martian winds are around 60 miles per hour, due to the lower density of the atmosphere. Martian winds move a lot of loose surface material, which is electrostatic, and can "sand blast" materiel. Giant Martian "dust devils" have been observed by the meteorological satellites in orbit. These swirling winds, similar to an Earth tornado, create large electrostatic fields, due to the dry conditions. Wind energy is fickle – you only get it when the wind blows.

Storage systems for that power, based on assemblies of highly efficient batteries are possible. Tesla is a leader in this area.

A combination of a large battery on the locomotive, and electrified rails or lines may be a possible solution. The locomotive motors and control system can be made quite efficient. The main job of the electrified lines would be to keep the battery charges, and the battery could be sized to allow the train to continue or return to its depot in a loss of power scenario. As a side effect, the weight of the battery will assist in traction.

The units could be autonomously operated, with a crew member in the cab for supervision.

A preferred arrangement would be to have all wheels driven, by their own AC electric motors. A control computer would adjust current and frequency to each motor on a sub-second basis, to avoid wheel slip, and maximize tractive power.

Rolling stock

There would be two major uses for the railroad. One would be to carry people and supplies from the landing site to the habitat area, or from one habitat to another. The launch/landing site would be separated from the habitat area for safety reasons.

Another use for the railroad would be to haul excavated material from the mining location, to the processing location. Ideally, you would build the processing facility next to the raw material, but this may not be possible or economical. What do we need on Mars that we might mine? Water ice comes to mind. Frozen methane is also useful. These need to go to an area with significant and consistent sunlight.

Rails and ballast

If we start on a traditional steel wheel on steel rail approach, these are questions and issues we need to address. Initially, anything we need has to brought form Earth, at great cost. Once we have a viable habitat, we will certainly go exploring for sources of materials that would limit the need for exports from Earth. The major needs are water, oxygen, and perhaps methane. We can bring with us infrastructure to process the raw materials into useful product. As was demonstrated in the Industrial Revolution on Earth, it is better to locate the processing facility close to the source of the raw material. This might be some distance from the ideal habitat and landing area. A cheap and reliable transportation method will be required. The railroad will be visible from orbit, and operations on the ground will get periodic updates from the meteorological satellites.

In terms of transportation, we can fly, use wheeled or tracked vehicles, or train technology. Flying is not much of an option in the lean atmosphere of Mars, and will probably be limited to drones. Large wheeled vehicles would need to be brought from Earth, as would the basis for any rail operations.

Since the primary scenario involves bring raw material from a source to a processing facility, we could consider a dedicated rail line. What type of rail? Are we talking steel rails on wooden sleepers (ties?). Probably not. Due to the decreased gravity, our train and its content will not weigh much, and we might get away with lighter aluminum rails, or plastic composite ties. A major problem in terrestrial desert rail lines is wind-blown dust. On Mars, the locomotive would have to have a plow, and perhaps a sweeper and brushes in front of all the wheels.

An alternative approach could be an overhead monorail system, like the one at DisneyWorld. This involves overhead structure, but light-weight materials such as aluminum or plastic could be used.

Another approach could be a MagLev design. These, unfortunately use a lot of power for levitation, and are probably not cost-effective, even on Earth, for heavy bulk traffic. In addition, wind-blown dust could interfere with the close clearances of the rail and the levitating body. One approach would be to encapsulate the rail operation in a long tube, elevated or buried. That adds to the cost and complexity of the project.

Elon Musk's hyperloop project is a maglev. It functions as a vactrain. All of the air in front of the train is pumped out, so the train, in a tube, moves in that direction. Keep in mind the air pressure at the Martian is about 0.6% of Earths. Probably, for the Martian vactrain, we would need to pump in atmosphere on the "push" end instead. The added infrastructure of the tube negates some of the advantages. It would have to be carried to Mars, or fabricated there of local materials. The track gauge could be a meter, or anything we choose.

If the rails are not made of metal, we can't use them to carry electric power to the motive unit. Electrified rails are dangerous. A lot of subway systems use a "third rail, shielded, to convey power. There is a trade-off here with overhead lines and centenary.

The right-of-way is prepared with a bulldozer, to level the ground and remove obstructions. Track laying can be done directly by machine, the technology is widely used in Earth applications. Tracks and ties can be laid in front of the track-laying machines, as they advance.

Fuel

Curiosity's exploration of the ancient lake bed, known as Gale Crater resulted in some new discoveries that NASA released on June 7, 2018. It found organic molecules, particularly methane, below the surface. Curiosity has a sampling drill (that, unfortunately, is limited to 5 cm.), a mass spectrometer, and a gas chromograph. On Earth, most methane is from biological processes. It can also be produced by inorganic processes, however. Scientists have also discovered a seasonal pattern in the amount of methane in the Martian atmosphere, by a factor of three, that may point to sub-surface storage.

The locomotive would not necessarily have a combustion engine. The fuel issue could be solved with indigenous methane, but the lack of oxygen is a problem. Oxygen would be reserved for breathing. If large amounts of water are found, some could be converted to hydrogen and oxygen by large solar array farms. That's fuel, right there, for a return trip home. Also, large greenhouses could be used to generate oxygen from carbon dioxide, but the infrastructure will take time to be developed. It might just be enough to balance the carbon dioxide from the crew quarters. If large deposits of carbon dioxide ice is found,the game changes.

Another potential fuel source is silane (SiH_4), which appears in the Martian regolith. There are processes that free up the hydrogen, and produce, as a waste product, semiconductor grade silicon. We don't even need to ship that back to Earth, we can run a semiconductor fab on Mars, and ship back commodity semiconductor chips

Further down the (rail) road, when multiple habitats are established, a system of trains, possible monorail or maglev, could be used as people-and light freight transportation between stations,

Maintenance-of-way

The Martian Railway will require a lot of maintenance and some repair. We will need to examine what types of work can be done in situ, on the rails, by a Martian-suited astronaut. Other more involved maintenance would involve bringing the vehicle inside a dedicated maintenance facility, fitted with airlocks. This could include more in-depth diagnostic equipment, and machine tools. Ideally, a 3-d print facility could fabricate parts as needed from specifications and drawings brought along, from material in inventory, or, later, derived from local sources.

Ideally, we would like the railroad infrastructure to last for a decade or more before major renovations are needed. The track may need periodic replacement, if we use aluminum or hardened plastic. The locomotives will need periodic shopping, and the main enemy will be dust in the machinery, leading to wear on moving parts. Dust in the bearings will be the major issue. A simple open car for passengers, requiring them to wear Mars-suits would be k, but at some point, we will want a closed and pressurized carriage, maybe with a bathroom. Other infrastructure, catenary, for example, may suffer wind erosion, and need periodic replacement.

Afterword

At some point after the first colony is established on Mars, we are going to have to move large amounts of cargo from point A to point B. As resupply ships come in from Earth, they will land at a safe distance fro the habitat, and we need to move the incoming supplies and new crew. If we discover a useful mining site, we will have to move something from the mine back to the workshops. In many areas, a wheeled vehicle works fine. For repetitive cargo moves, a train is ideal.

The Martian environment differs quite a bit from Earth's, in terms of atmospheric composition and pressure, gravity, surface, winds, etc. The technology of railroads will have to be adjusted to work in the Martian environment. That is left as an exercise for the student. When I get there, my train ticket had better be waiting.

Bibliography

Aldrin, Buzz; David, Leonard *Mission to Mars: My Vision for Space Exploration,* 2015, ISBN-1426214685.

Aldrin, Buzz; Dyson, Marianne *Welcome to Mars: Making a Home on the Red Planet,* 2015, ISBN-1426322062.

Baker, David *NASA Mars Rovers Manual: 1997-2013 (Sojourner, Spirit, Opportunity and Curiosity) (Owners' Workshop Manual), 2013,* ISBN-0857333704.

Bell, Jim *Postcards from Mars: The First Photographer on the Red Planet,* Dutton Adult, 2016, ISBN-0525949852.

Bradbury, Ray *The Martian Chronicles,* ISBN-006207993X.

Burns, Jack O,   et al, NASA Ames, "SCIENCE AND EXPLORATION AT THE MOON AND MARS ENABLED BY SURFACE" TELEROBOTICS" INTERNATIONAL ACADEMY OF ASTRONAUTICS 10th IAA SYMPOSIUM ON THE FUTURE OF SPACE EXPLORATION: TOWARDS THE MOON VILLAGE AND BEYOND, Torino, Italy, June 27-29, 2017. avail: https://www.lpi.usra.edu/science/kring/lunar_exploration/BurnsEt Al_IAA-2017_SurfaceTelerobotics.pdf

Burt, Dennis *Elon Musk will Take Us to Mars: How and Why the Billionaire Entrepreneur and his SpaceX Start-Up are Making Interplanetary Travel a Reality,* 2013, ASIN-B00G0T6D6U.

Carney, Elizabeth *Mars, The Red Planet,* National Geographic Kids, ISBN-978-1-4263-2754-4.

Chandra, Satish. Railway Engineering, 2013, Oxford University

Press, ISBN-9780198083535.

Charles River Editors, *The Transcontinental Railroad: The History and Legacy of the First Rail Line Spanning the United States*, 2014, ASIN-B00QKRNJ4Q.

Cichana, Timothy; et al, "MARS BASE CAMP UPDATES AND NEW CONCEPTS," 2017, IAC-17, 68th International Astronautical Congress (IAC).

Cichana, Timothy; et al, "Mars Base Camp: An Architecture for Sending Humans to Mars," 2017, New Space, Vol 5, No 4, avail: https://www.liebertpub.com/doi/full/10.1089/space.2017.0037.

David, Leonard *Mas: Our Future on the Red Planet,* 2016, ISBN-426217587.

Devaney, Sherri *The Qinghai-Tibet Railway,* A Great Idea: Engineering, 2013, ISBN-9781603575799.

Editors of National Geographic; Daniels, Patricia *National Geographic Mars: Secrets of the Red Planet,* 2018, ISBN-1547842466.

Harland, David M. *Mars Owners' Workshop Manual: From 4.5 billion years ago to the present (Haynes Manuals),* ISBN-1785211382.

Hohmann, Walter *THE ATTAINABILITY OF HEAVENLY BODIES,* Reprinted as NASA TT-F-44, 2017.
avail:
http://large.stanford.edu/courses/2014/ph240/nagaraj1/docs/hohmann.pdf

Housley, Kathleen L. *Fire and Forge: A Desert Railroad, a Wonder Metal, and the Making of an Aerospace Blacksmith, 2013,*

ASIN-B079J5HHDJ.

Kaufman, Mark *Mars Up Close: Inside the Curiosity Mission*, 2014 National Geographic, ISBN-142621278X.

Lakdawalla, Emily *The Design and Engineering of Curiosity: How the Mars Rover Performs Its Job*, 2018, Springer, ISBN-3319681443

Ley, Willy and Von Braun, Werner, *The Exploration of Mars*, 1956, Sidgwick & Jackson; First Edition, ASIN-B0000CJKQN.

Lowell, Perciva, *Mars and its Canals*, ASIN-B00U3ZRER4.

Lowell, Percival, *Mars: Is There Life On Mars?*, ISBN-1605065528.

Lowell, Percival, *Mars*, ASIN-B004ISKE6K.

Lowell, Percival, *The Evolution of Worlds*, ASIN-B00MSC2PJW.

Lowell, Percival, *Mars as the abode of life*, ASIN-B00JG57GA2.

Mackenzie, Bruce "To Mars - a Permanent Settlement on the First Mission," presented at the 1998 International Space Development Conference, May 21–25, Milwaukee WI.

McGinty, Michael *Canals on Mars: Lowell Revisited*, 2016, ASIN-B01BCD66BK.

NASA, *NASA's Constellation Program: Lessons Learned (Volume I and II) - Moon and Mars Exploration Program - Ares Rockets and Orion Spacecraft, avail:*
http://www.thebookishblog.com/nasa-s-constellation-program-lessons-learned-volume-i-and-ii.pdf

NASA, *NASA Report on Mars Exploration: Frontier In-Situ*

Resource Utilization for Enabling Sustained Human Presence on Mars - ISRU, Surface Habitats, Entry Descent and Landing, Fuels, Food, Robotics, 2016, ASIN-B01JDORX58.

NASA, *NASA Space Technology Report: Lunar and Planetary Bases, Habitats, and Colonies, Special Bibliography Including Mars Settlements, Materials, Life Support, Logistics, Robotic Systems,* ASIN-B00CLX44E2.

NASA, *NASA Space Technology Report: Deep Space Habitat Concept of Operations for Transit Mission Phases - Mars, Phobos / Deimos, Near Earth Asteroid, Habitats, Crew Systems,* 2013, ASIN-B00EG4N3E6.

National Geographic, "Mars: Inside the High-Risk, High Stakes Race to the Red Planet, November 2016, https://www.nationalgeographic.com/magazine/2016/11/#

Paris, Antonio *Mars: Your Personal 3D Journey to the Red Planet,* 2018, The Center for Planetary Science, ISBN-0692073671.

Petranek, Stephen *How We'll Live on Mars*, 2015, Simon & Schuster, ISBN-1476784760.

Portree, David S. F. *Humans to Mars: Fifty Years of Mission Planning, 1950–2000*, NASA Monographs in Aerospace History Series, Number 21, February 2001, NASA SP-2001-4521. Avail: ASIN-B014RGH7GM.

Rapp. Donald *Human Missions to Mars: Enabling Technologies for Exploring the Red Planet,* 2015, Springer Praxis, ISBN-3319222481.

Sparrow, Giles *MARS* (Illustrated), 2015, ISBN-162365856X.

Stakem, Patrick H. *Mars*, 2018, PRRB Publishing, ISBN-1983116904.

Tolker-Nielsen, Toni "Exomars 2016 – Schiaparelli Anomaly Inquiry," 2017, ESA, DG-I/2017/546/TTN.

U. S. Department of Defense, *Railway Track Maintenance*, 2009, ASIN-B0026RI430.

Valier, Max; Miller, Ron (ed), *A Daring Trip to Mars*,1928, reprint, 2013, ASIN-B00CSWANK0.

Von Braun, Werhner *Project MARS, a Technical Tale*, 1971, ISBN-0-9738203-3-0.

Von Braun, Werhner *The Mars Project*, 1962, U. Illinois Press, ISBN-0252062272.

Von Braun, Wernher *Project MARS: A Technical Tale, 2006*, ISBN-0973820330.

Wallace, Alfred Russel *Is Mars habitable? A critical examination of Professor Percival Lowell's book "Mars and its canals," with an alternative explanation*, ASIN-B0084C22WK.

Webb, Walter Loring *Railroad Construction - Theory And Practice - A Textbook For The Use Of Students In Colleges And Technical Schools*, reprint 2008, ISBN-1408640244.

Weintraub, David *Life on Mars: What to Know Before We Go*, 2018, ASIN-B079P8S585.

Wolmar, Christian *To the Edge of the World: The Story of the Trans-Siberian Express, the World's Greatest Railroad*, 2014, ASIN-B00J1JPSRQ.

Zubrin, Robert *Mars on Earth: The Adventures of Space Pioneers in the High Arctic*, 2003, ISBN-158542255X .

Zubrin, Robert *Entering Space: Creating a Spacefaring Civilization*, 2000, ISBN-10-1585420360.

Zubrin, Robert *Mars Direct: Space Exploration, the Red Planet, and the Human Future: A Special from Tarcher/ Penguin,* 2013, ASIN-B00AMOO98I.

Resources

- http://newmars.com/forums/viewtopic.php?id=3501&p=6
- UP - http://trn.trains.com/news/news-wire/2016/09/29-union-pacific-on-mars
- http://theconversation.com/elon-musks-high-speed-hyperloop-train-makes-more-sense-for-mars-than-california-43686
- http://cs.trains.com/mrr/f/88/t/16250.aspx – model RR
- https://www.reddit.com/r/SurvivingMars/comments/9z2qhy/feature_request_monorail_long_distance_dome_to/
- https://futurism.media/space-exploration-developments-by-2050-a-fictional-vision A. C. Clarke
- Human Exploration of Mars, Reference Mission avail: https://web.archive.org/web/20070626154441
- http://exploration.jsc.nasa.gov/marsref/contents.html
- Maps are available at Google Mars.
- http://www.nasa.gov/mars
- Mars Base Camp, http://lockheedmartin.com/us/ssc/mars-orion.html
- http://cs.trains.com/mrr/f/88/t/16250.aspx
- NASA, Systems Engineering Handbook: NASA/SP-2016-6105 Rev2, 2017. avail:
- https://www.nasa.gov/connect/ebooks/nasa-systems-engineering-handbook
- http://theconversation.com/elon-musks-high-speed-hyperloop-train-makes-more-sense-for-mars-than-california-43686.
- A. C. Clarke - https://futurism.media/space-exploration-developments-by-2050-a-fictional-vision
- Atreya, Sushil K.; Mahaffy, Paul R.; Wong, Ah-San (2007). "Methane and related trace species on Mars: origin, loss, implications for life, and habitability".Planetary and Space Science.55(3):, avail:

https://www.sciencedirect.com/science/article/pii/S0032063306001

814?via%3Dihub
- wikipedia, various.

References – Mars Exploration rover

https://mars.jpl.nasa.gov/mer/home/index.html

References – Mars Global surveyor

mars.jpl.nasa.gov/mgs/

References – Mars Exploration rover

https://mars.jpl.nasa.gov/mer/home/index.html

Glossary of Terms

Areography – equivalent of geography on Earth.

Apogee – furthest point in the orbit from the Earth.

Apoareion – farthest point in orbit, from Mars.

ASIN – Amazon Standard Inventory Number

Astrionics – electronics for space flight.

AU – astronomical unit.

Baud – measure of data rate; bits per second

BEM – bug-eyed monster

BEO – beyond Earth orbit.

Byte – data structure of 8 bits.

CBM – common berthing mechanism.

CCS – command computer system

CCSDS – Consultive Committee on Space Data Systems, a standards organization.

CM – crew module

CME – Coronal Mass Ejection, blast of energetic particles from the Sun.

CMP – co-manifested payload.

CNSA – China National Space Administration.

Conops – concept of operations.

CPS – Cyrogenic Propulsion Stage.

CPU – central processing unit

CRTBP – Circular Restricted three-body Problem.

CSA – Canadian Space Agency, Agence Spatiale Canadienne

CSF – Cislunar Support Flight.

C&W – caution and warning.

Cygnus – Orbital-ATK automated cargo vehicle for ISS.

Cyrogenic – relating to very low temperatures.

DAM – damage avoidance maneuver.

DCM – docking cargo module.

Delta-V – change in velocity.

DoD – (U.S.) Department of Defense

DRG – Distant Retrograde Orbit.

DRM – design reference mission.

DSG – Deep Space Gateway

DSH – deep space habitat.

DSN – (NASA) Deep Space Network.

DST – Deep Space Transport

DTM – dynamic test model, for structural tests.

ECLSS – Environmental Control & Life Support system.

EDL – Entry, Descent, Landing.

EMD – Electromotive Division,a U. S. locomotive manufacturer.

EM-x Exploration Mission number-x.

EMM – Emirates Mars Mission.

Ephemeris – position information data set for orbiting bodies, 6 parameters plus time.

Epoch – a reference point in time for orbital elements.

EPS – electrical power system

ESA – European Space Agency

EUS – Exploration Upper Stage.

EVA – extra-vehicular activity.

Flash – a type of non-volatile memory

FMARS – Flashline Mars Arctic Research Station

FPGA – Field Programmable Gate Array – an integrated circuit

FTL – faster than light

FTP – file transfer protocol.

G – one Earth normal gravity; as a prefix, 10^{12}

Gauge – distance between the rails.

GCSC – Guidance, control, sequencing.

GHZ - giga-Hertz

GNC – Guidance, Navigation, and Control.

Gravity well – a conceptual model of the gravity field near a mass.

GSFC – NASA Goddard Space Flight Center, Greenbelt, MD.

GYRO – sensor for orientation.

Halo Orbit – three dimension orbit near the L1, L2, or L3 Lagrange points.

HEEO – highly eccentric Earth orbit.

HEOMD – Human Exploration and Operations Mission Directorate.

HITL – Human in the loop.

HOPE – Human Outer Planet Exploration (NASA)

HSIR – human systems integration requirements
IDSS – International Docking System Standard.
IGA - (ISS) InterGovernmental Agreement
IMU – inertial measurement unit.
IP – internet protocol
ISP – specific impulse. Measure of efficiency of rocket engine.
    Units of seconds.
ISRO – Indian Space Research Organization
ISRU – in situ resource utilization
ISS – International Space Station
JAXA – Japan Aerospace Exploration Agency.
KW – kilowatt.
IDSN – Indian deep space network
ISRU – in site resource utilization.
ISS – International Space Station
JAXA – Japanese space agency
JPL – Jet Propulsion Laboratory, Pasadena, CA.
JSC – Johnson Space Center, Houston, Texas.
K, kilo - 10^3
KSC – NASA Kennedy Space Center, launch site, Florida.
L2 – second of 5 Lagrange points, a null in the gravity field in the
    restricted 3-body problem.
LAS – launch abort system
Lbf – pounds, force.
LCT – Lunar Cargo Transportation.
LEO – Low Earth Orbit.
LGM – little green men.
LH2 – liquid hydrogen.
Libration point – null in the gravity field of the three body
    problem.
LOS – Russian Lunar Orbital Station; loss-of-signal.
LOX – liquid oxygen, boils at -297 F.
LSAM – lunar surface access module
LSPPO – Lunar Surface systems Project Office (NASA-JSC).
LST – landing by soft touchdown.
MADV – Mars Ascent/Descent Vehicle.
MBC – Mars Base Camp.

MER – Mars Exploration rover

MET – mission elapsed time.

MHz – mega (10^6) hertz

MIPS – millions of instructions per second.

MMSEV – MultiMission Space Exploration Vehicle.

MOU – memorandum of understanding.

MPCV - Multi-Purpose Crew Vehicle.

MPH – miles per hour

MPLM – Multi-purpose Logistics Module.

MRO – Mars Reconnaissance Orbiter

m/s – meters per second.

MSL – Mars Science Laboratory

MT – metric ton, 1000 kg.

N – Newton, metric unit of force.

NAC – NASA Advisory Council.

Nadir – the point directly below.

NASA – (U.S.) National Aeronautics and Space Administration

NEO – near Earth object.

NextSTEP-2 – (NASA) Next Space Technologies of Exploration Partnerships.

NHV – net habitable volume.

NRHO – Near rectilinear halo orbit (around the L1 or L2 Earth-Moon libration point).

NTIS – National Technical Information Service (www.ntis.gov).

NTRS – NASA Technical Reports Server, ntrs.nasa.gov

ORU – Orbital Replacement Unit.

OPSEK – (Russian) Orbital Piloted Assembly and Experiment Complex.

OWLT – one-way light time.

RAD – unit of radiation

Perigee –closest point in the orbit from the Earth.

Periareion – closest point in orbit to Mars.

PMA – Pressurized mating adapter.

PMCU – Power Management Control Unit.

PPB – power and propulsion bus.

PPO – Planetary protection officer.

PTCS – Passive thermal control system

PVCU – Photo Voltaic Control Unit.

R&D – research & development.

RAM – random access memory

RCS – reaction control system.

Regolith – layer of loose material, covering rock; dirt.

RGA – rate gyro assembly

ROSCOSMOS – Russian Space Agency.

RPOD – Rendezvous, Proximity Operations, Docking.

RTG – Radioisotope Thermoelectric Generator.

SEP – solar electric propulsion.

SEU – single event upset, transient error in a digital circuit, usually due to radiation.

SHFE – space human factors engineering.

SI – System International – the metric system.

Sidereal period – time for an object to make a full orbit.

Sol, local solar day – on Mars, 24h, 37 min.

SLS – (NASA) Space Launch System.

SPACE Act - Spurring Private Aerospace Competitiveness and Entrepreneurship

SPACE-X – private space company.

STEM - science, technology, engineering, math.

STEAM - science, technology, engineering, art, math.

Synodic period - time for an object in orbit to occupy the same point, in relation to 2 other objects.

TCS – thermal control system.

Tera – 10^{12}

TLI – Trans-lunar injection.

TM – Technical Manual.

TPS – thermal protection system.

Trillion - 10^{12}

TRL – technology readiness level.

UAE – United Arab Emirates

UDM – universal docking module.

UHF – ultra high frequency, 300 MHz to 3 Ghz

Ullage – residual fuel or oxidizer in a tank after engine burn is complete.

USAF – United States Air Force.

USGS - United States Geologic Survey.
UV – ultraviolet light.
V&V – verification and validation.
WDV – water delivery vehicle.
X-band – 8 – 12 GHz.
XBASE - Expandable Bigelow Advanced Station Enhancement.
Zenith – the point directly above.
Zombie-sat – a non functional satellite in orbit, contributing to the
    orbital debris problem.

If you enjoyed this book, you might also be interested in some of these from the author.

Stakem, Patrick H. *16-bit Microprocessors, History and Architecture*, 2013 PRRB Publishing, ISBN-1520210922.

Stakem, Patrick H. *4- and 8-bit Microprocessors, Architecture and History*, 2013, PRRB Publishing, ISBN-152021572X,

Stakem, Patrick H. *Apollo's Computers,* 2014, PRRB Publishing, ISBN-1520215800.

Stakem, Patrick H. *The Architecture and Applications of the ARM Microprocessors,* 2013, PRRB Publishing, ISBN-1520215843.

Stakem, Patrick H. *Earth Rovers: for Exploration and Environmental Monitoring,* 2014, PRRB Publishing, ISBN-152021586X.

Stakem, Patrick H. *Embedded Computer Systems, Volume 1, Introduction and Architecture*, 2013, PRRB Publishing, ISBN-1520215959.

Stakem, Patrick H. *The History of Spacecraft Computers from the V-2 to the Space Station*, 2013, PRRB Publishing, ISBN-1520216181.

Stakem, Patrick H. *Floating Point Computation*, 2013, PRRB Publishing, ISBN-152021619X.

Stakem, Patrick H. *Architecture of Massively Parallel Microprocessor Systems*, 2011, PRRB Publishing, ISBN-1520250061.

Stakem, Patrick H. *Multicore Computer Architecture,* 2014, PRRB Publishing, ISBN-1520241372.

Stakem, Patrick H. *Personal Robots*, 2014, PRRB Publishing, ISBN-1520216254.

Stakem, Patrick H. *RISC Microprocessors, History and Overview,* 2013, PRRB Publishing, ISBN-1520216289.

Stakem, Patrick H. *Robots and Telerobots in Space Applications*, 2011, PRRB Publishing, ISBN-1520210361.

Stakem, Patrick H. *The Saturn Rocket and the Pegasus Missions, 1965,* 2013, PRRB Publishing, ISBN-1520209916.

Stakem, Patrick H. *Microprocessors in Space*, 2011, PRRB Publishing, ISBN-1520216343.

Stakem, Patrick H. Computer *Virtualization and the Cloud*, 2013, PRRB Publishing, ISBN-152021636X.

Stakem, Patrick H. *What's the Worst That Could Happen? Bad Assumptions, Ignorance, Failures and Screw-ups in Engineering Projects, 2014,* PRRB Publishing, ISBN-1520207166.

Stakem, Patrick H. *Computer Architecture & Programming of the Intel x86 Family, 2013,* PRRB Publishing, ISBN-1520263724.

Stakem, Patrick H. *The Hardware and Software Architecture of the Transputer,* 2011,PRRB Publishing, ISBN-152020681X.

Stakem, Patrick H. *Mainframes, Computing on Big Iron*, 2015, PRRB Publishing, ISBN- 1520216459.

Stakem, Patrick H. *Spacecraft Control Centers*, 2015, PRRB Publishing, ISBN-1520200617.

Stakem, Patrick H. *Embedded in Space,* 2015, PRRB Publishing, ISBN-1520215916.

Stakem, Patrick H. *A Practitioner's Guide to RISC Microprocessor Architecture*, Wiley-Interscience, 1996, ISBN 0471130184.

Stakem, Patrick H. *Graphics Processing Units, an overview*, 2017, PRRB Publishing, ISBN-1520879695.

Stakem, Patrick H. *Intel Embedded and the Arduino-101, 2017,* PRRB Publishing, ISBN-1520879296.

Stakem, Patrick H. *Orbital Debris, the problem and the mitigation*, 2018, PRRB Publishing, ISBN-*1980466483*.

Stakem, Patrick H. *Manufacturing in Space*, 2018, PRRB Publishing, ISBN-1977076041.

Stakem, Patrick H., *NASA's Ships and Planes*, 2018, PRRB Publishing, ISBN-1977076823.

Stakem, Patrick H. *Space Tourism*, 2018, PRRB Publishing, ISBN-1977073506.

Stakem, Patrick H. *STEM – Data Storage and Communications*, 2018, PRRB Publishing, ISBN-1977073115.

Stakem, Patrick H. *In-Space Robotic Repair and Servicing*, 2018, PRRB Publishing, ISBN-1980478236.

Stakem, Patrick H. *Introducing Weather in the pre-K to 12 Curricula, A Resource Guide for Educators*, 2017, PRRB Publishing, ISBN-1980638241.

Stakem, Patrick H. *Introducing Astronomy in the pre-K to 12 Curricula, A Resource Guide for Educators*, 2017, PRRB Publishing, ISBN-198104065X.

Also available in a Brazilian Portguese edition, ISBN-1983106127.

Stakem, Patrick H. *Deep Space Gateways, the Moon and Beyond*, 2017, PRRB Publishing, ISBN-1973465701.

Stakem, Patrick H. *Crewed Spacecraft*, 2017, PRRB Publishing, ISBN-1549992406.

Stakem, Patrick H. *Rocketplanes to Spacecraft*, 2017, PRRB Publishing, ISBN-1549992589.

Stakem, Patrick H. *Crewed Space Stations*, 2017, PRRB Publishing, ISBN-1549992228.

Stakem, Patrick H. *,Enviro-bots for STEM: Using Robotics in the pre-K to 12 Curricula, A Resource Guide for Educators*, 2017, PRRB Publishing, ISBN-1549656619.

Stakem, Patrick H. *STEM-Sat, Using Cubesats in the pre-K to 12 Curricula, A Resource Guide for Educators*, 2017, ISBN-1549656376.

Stakem, Patrick H. *Visiting the NASA Centers, and Locations of Historic Rockets and Spacecraft*, 2107, PRRB Publishing, ISBN-154965120X.

Stakem, Patrick H. *Lunar Orbital Platform-Gateway*, 2018, PRRB Publishing, ISBN-1980498628.

Stakem, Patrick H. Embedded GPU's, 2018, PRRB Publishing, ISBN- 1980476497.

Stakem, Patrick H. *Mobile Cloud Robotics*, 2018, PRRB Publishing, ISBN- 1980488088

Stakem, Patrick H. *Extreme Environment Embedded Systems* 2017,

PRRB Publishing, ISBN-1520215967.

Stakem, Patrick H. *What's the Worst, Volume-2*, 2018, ISBN-1981005579.

Stakem, Patrick H., *Spaceports*, 2018, ISBN-1981022287.

Stakem, Patrick H., *Space Launch Vehicles*, 2018, ISBN-1983071773.

Stakem, Patrick H. *Mars*, 2018, ISBN-1983116902.

Stakem, Patrick H. *X-86, 40th Anniversary ed*, 2018, ISBN-1983189405.

Stakem, Patrick H. *Exploration of the Gas Giants and the Ice Giants, Space Missions to Jupiter, Saturn, Uranus, and Neptune*, 2017, ISBN-1717814506.

Stakem, Patrick H. Rocket Science-101, 2018, ISBN-1977067697.

Stakem, Patrick H. *Lunar Orbiting Platform-Gateway*, 2017, ISBN-1980498628.

Stakem, Patrick H. *Space Weathe*r, 2018, ISBN-1723904023.

Stakem, Patrick H. *STEM-Engineering Process*, 2017, ISBN-1983196517.

Stakem, Patrick H. *RISC-V in Space*, 2019, ISBN-1796434388 .

Stakem, Patrick H. *Mars Railroad,* 2019, ISBN–1794488243.

Stakem, Patrick H. *Arm in Space*, 2019, ISBN-1099789133.

Stakem, Patrick H. *Exploiting the Moon,* ISBN-978-1091057852.

Stakem, Patrick H. Terraforming, 2018, ISBN-978-1790308101.

Stakem, Patrick H. *Exploration of the Asteroid Belt, a new approach*, 2018, ISBN-978-1731049841.

Stakem, Patrick H. *Exoplanets,* 2018, ISBN-978-1731385055.

Stakem, Patrick H. *Planetary Defense*, 2018, ISBN-978-1731001207.

Stakem, Patrick H. *Space Telescopes*, 2018, ISBN-978-1728728568.

Robotic Exploration of the Icy moons of the Gas Giants, 2020, ISBN-979-8621431006.

Stakem, Patrick H. *Riverine Ironclads, Submarines, and Aircraft Carriers of the American Civil War*, 2019, PRRB Publishing.

Stakem, Patrick H. *Submarine Launched Ballistic missiles,* ISBN-978-1088954904.

Stakem, Patrick H., *Space Command, Military in Space,* ISBN-978-1693005398.

Cubesat Engineering, 2017, ISBN-1520754019.

Cubesat Operations, 2017, ISBN-152076717X.

Interplanetary Cubesats, 2017, ISBN-1520766173 .

Cubesat Constellations, Clusters, and Swarms, 2017, ISBN-1520767544.

Solar Sailing for Cubesats, 2020, ISBN-979-8634968650.

Hacking Cubesats, Cybersecurity in Space, 2020, ISBN-979-

8623458964.

History & Future of Cubesats, 2020, ISBN-979-8649179386 .

CubeRovers, A Synnergy of Technology, 2020, ISBN-979-8651773138.

Deep Space Cubesats, 2020, ISBN-9798654272409.

Exploration of Lunar & Martian Lava Tubes by Cube-X, 2020, ISBN-979-8621435325.

www.ingramcontent.com/pod-product-compliance
Lightning Source LLC
Chambersburg PA
CBHW030537220526
45463CB00007B/2878